LET'S-READ-AND-FIND-OUT SCIENCE® STAGE 2

Who Eats What?

Food Chains and Food Webs

by Patricia Lauber • illustrated by Holly Keller

HarperCollins*Publishers*

The illustrations in this book were done with watercolor and black pen on Fabriano paper.

The *Let's-Read-and-Find-Out Science* book series was originated by Dr. Franklyn M. Branley, Astronomer Emeritus and former Chairman of the American Museum–Hayden Planetarium, and was formerly co-edited by him and Dr. Roma Gans, Professor Emeritus of Childhood Education, Teachers College, Columbia University. Text and illustrations for each book in the series are checked for accuracy by an expert in the relevant field. For a complete catalog of Let's-Read-and-Find-Out Science books, write to HarperCollins Children's Books, 10 East 53rd Street, New York, NY 10022.

Let's Read-and-Find-Out Science is a registered trademark of HarperCollins Publishers.

Library of Congress Cataloging-in-Publication Data
Lauber, Patricia.
  Who eats what?: Food chains and food webs / by Patricia Lauber ; illustrated by Holly Keller.
    p.   cm. — (Let's-read-and-find-out science. Stage 2)
  Summary: Explains the concept of a food chain and how plants, animals, and humans are ecologically linked.
  ISBN 0-06-022981-0. — ISBN 0-06-022982-9 (lib. bdg.). — ISBN 0-06-445130-5 (pbk.)
  1. Food chains (Ecology)—Juvenile literature.   [1. Food chains (Ecology).   2. Ecology.]   I. Keller, Holly, ill.   II. Title.   III. Series.
QH541.14.L38   1995   93-10609
574.5'3—dc20   CIP
                                                                                                    AC

Typography by Elynn Cohen
1   2   3   4   5   6   7   8   9   10
❖
First Edition

Who Eats What?

A caterpillar is eating a leaf on an apple tree.

Later the caterpillar is spotted by a wren.
It becomes part of the wren's dinner.

Still later the wren is eaten by a hawk.

Leaf, caterpillar, wren, and hawk are all linked. Together they form a food chain. Each is a link in the chain.

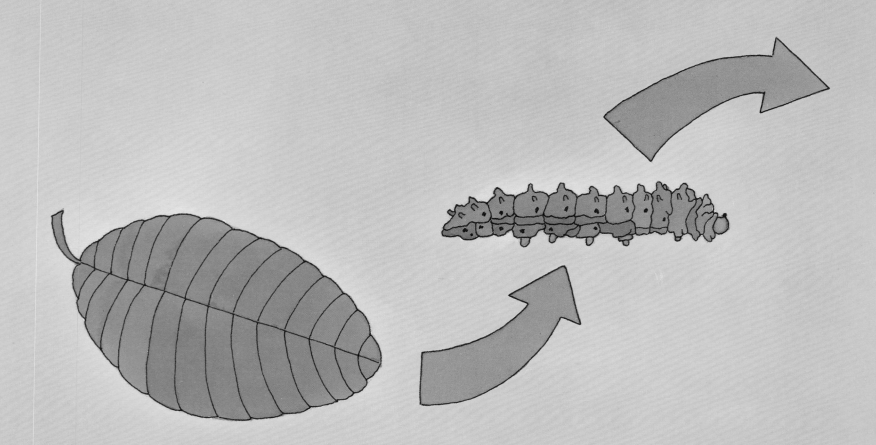

The hawk is the top of the food chain, because no other animal attacks and eats hawks. The animal at the top of a food chain is always the last eater—the one nobody else eats.

Suppose you eat an apple off the tree. That makes you part of a short food chain—the apple and you. You are the top of the food chain.

Or suppose you drink a glass of milk. Now you are the top of a slightly longer food chain. The milk came from a cow, and the cow ate grass. So this chain is grass, cow, you.

Every time you eat a meal, you become the top of several food chains. You can draw a picture to show them. If you had a peanut-butter-and-jelly sandwich, a glass of milk, and an apple, the picture might look like this.

milk   peanut butter   grape jelly   bread   apple

Food is the fuel our bodies need. Food keeps us alive. It gives us the energy we need to grow, move, and do many other things. The same thing is true for caterpillars, wrens, hawks—for all animals. All must find or catch the foods they need.

When you draw a food chain, you are drawing a flow of energy. The arrows show its path.

There are many, many food chains, more than anyone can count. But in one way they are all alike.

All food chains begin with green plants. Green plants are the only living things that can make their own food. They are the only living things that do not need to eat something else.

Green plants take energy from sunlight. They use it to make food out of water and air.

All animals depend on green plants for food, even animals that don't eat plants.

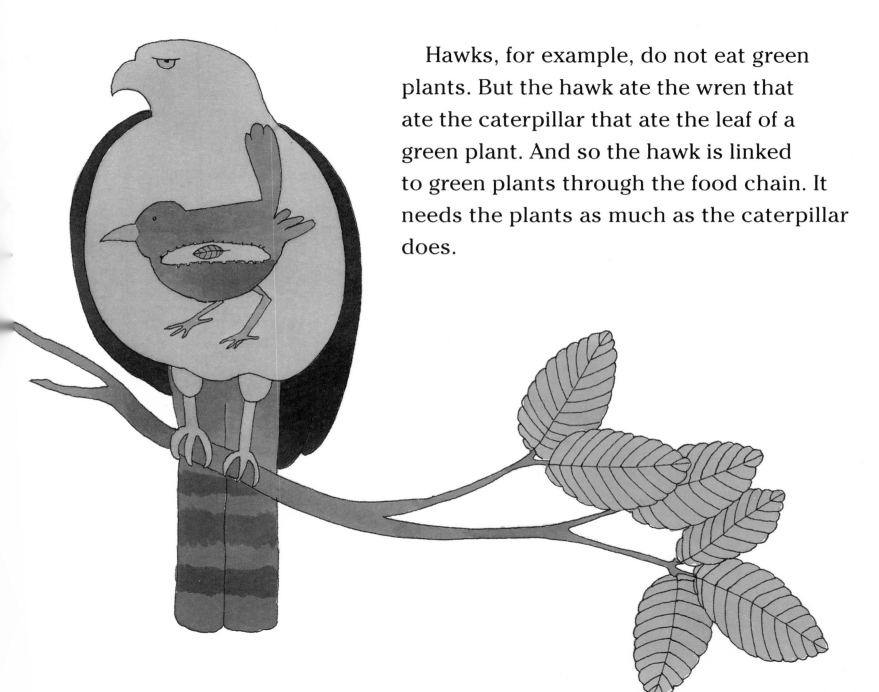

Hawks, for example, do not eat green plants. But the hawk ate the wren that ate the caterpillar that ate the leaf of a green plant. And so the hawk is linked to green plants through the food chain. It needs the plants as much as the caterpillar does.

Take a walk and look around. You will see parts of many food chains. Look at the leaves and flowers of plants. Look at the bark of trees. Look at fruits, nuts, and seeds that have fallen to the ground. What animals are eating them?

You might see a grasshopper eating a blade of grass. You may not see another animal eat the grasshopper. But you can find out which animals eat grasshoppers by going to the library. You can read up on grasshoppers and any other animal you've seen. You can draw food chains.

15

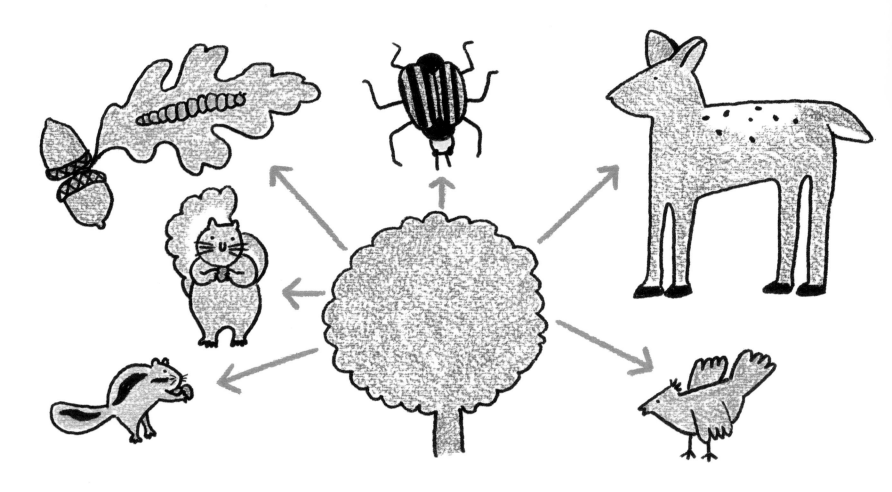

Your drawings will show that one plant may be the start of several food chains. The leaves of an oak tree may be food for caterpillars. Beetles may bore into the tree's trunk. Acorns are food for squirrels, chipmunks, blue jays, and deer.

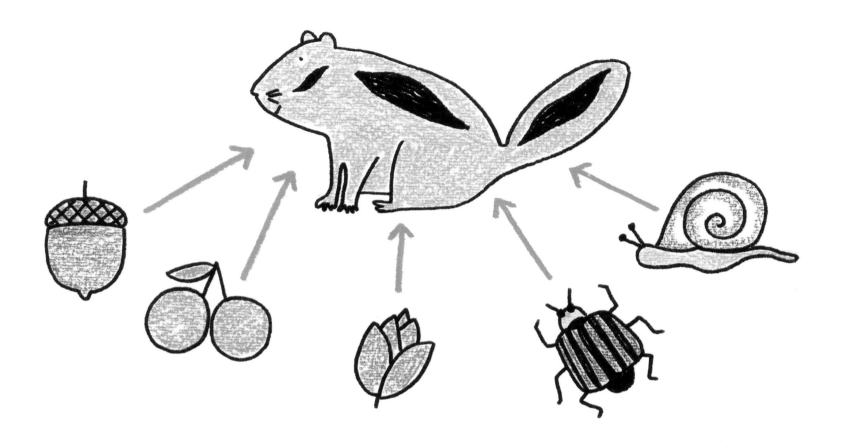

The drawings will also show that most animals are part of several food chains. Chipmunks, for example, eat many foods. They eat nuts, seeds, berries, buds. They may also eat insects, snails, and other small animals.

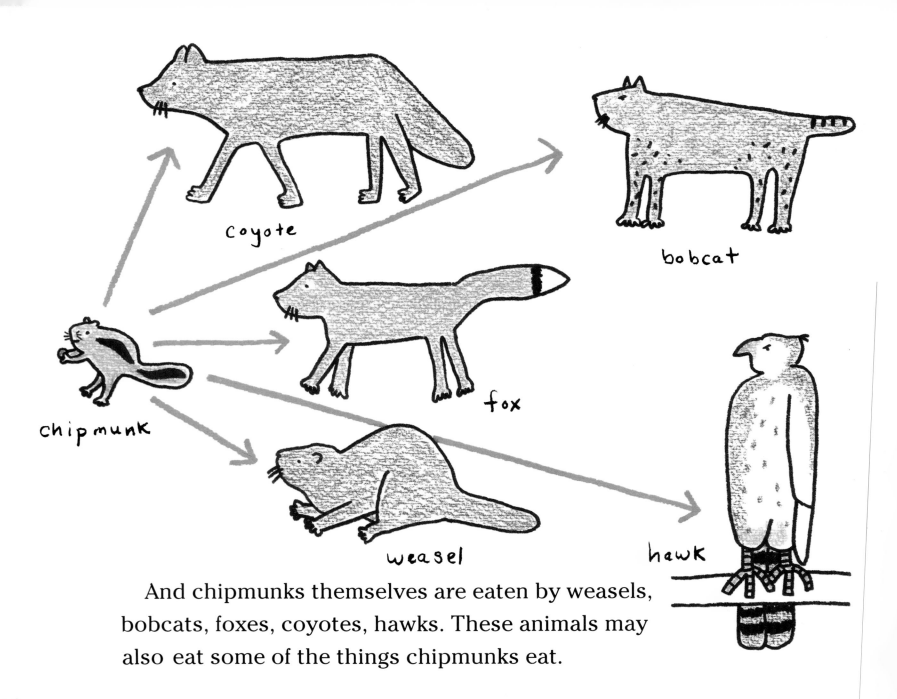

And chipmunks themselves are eaten by weasels, bobcats, foxes, coyotes, hawks. These animals may also eat some of the things chipmunks eat.

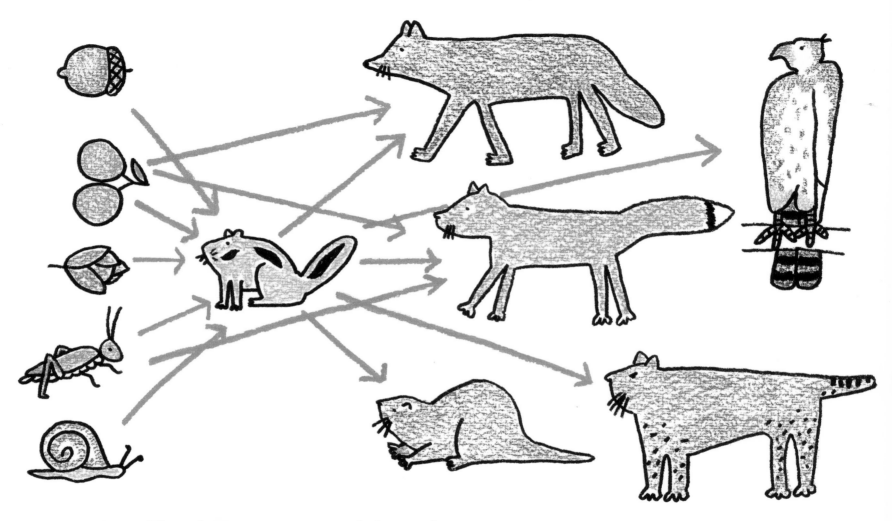

Try drawing some of these food chains on one page. You will have arrows branching in all directions. Now you have drawn a food web. Food webs are made up of many food chains.

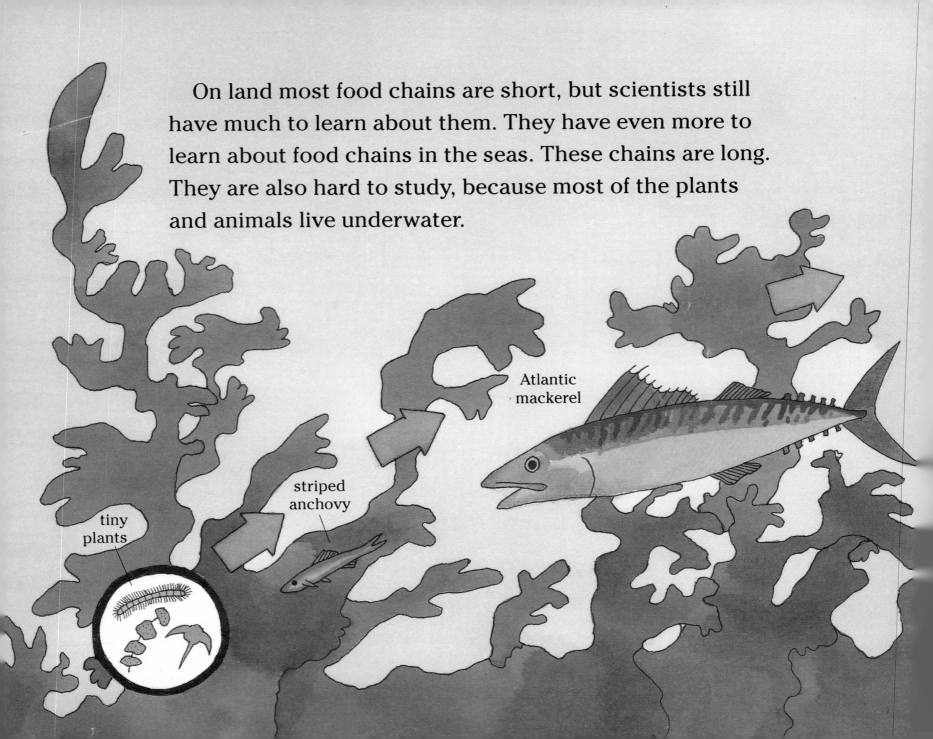

On land most food chains are short, but scientists still have much to learn about them. They have even more to learn about food chains in the seas. These chains are long. They are also hard to study, because most of the plants and animals live underwater.

Atlantic mackerel

striped anchovy

tiny plants

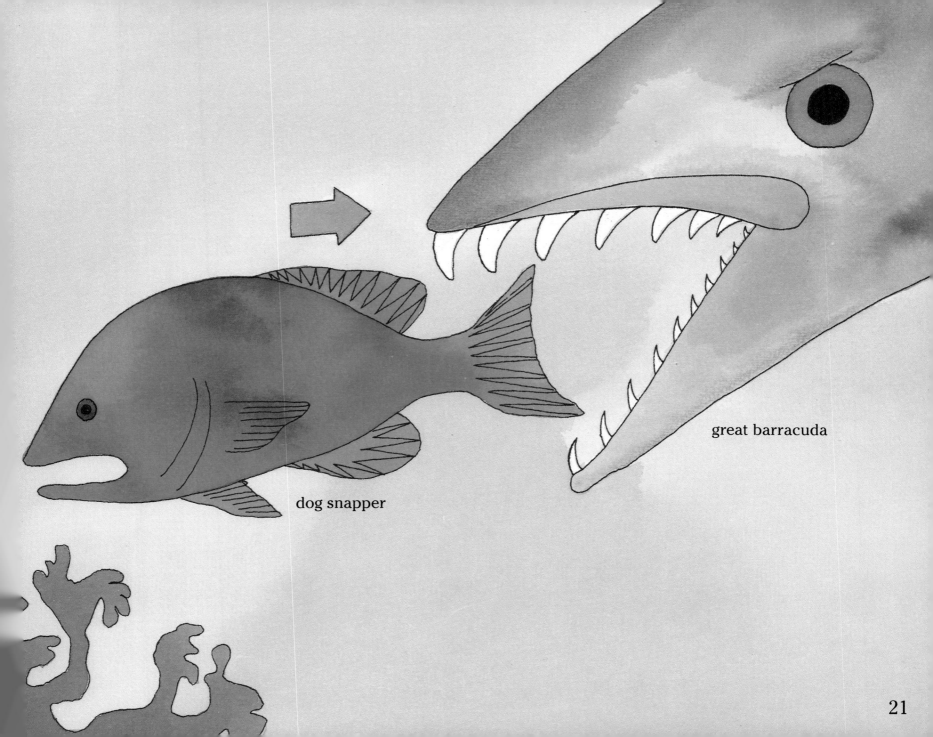

dog snapper

great barracuda

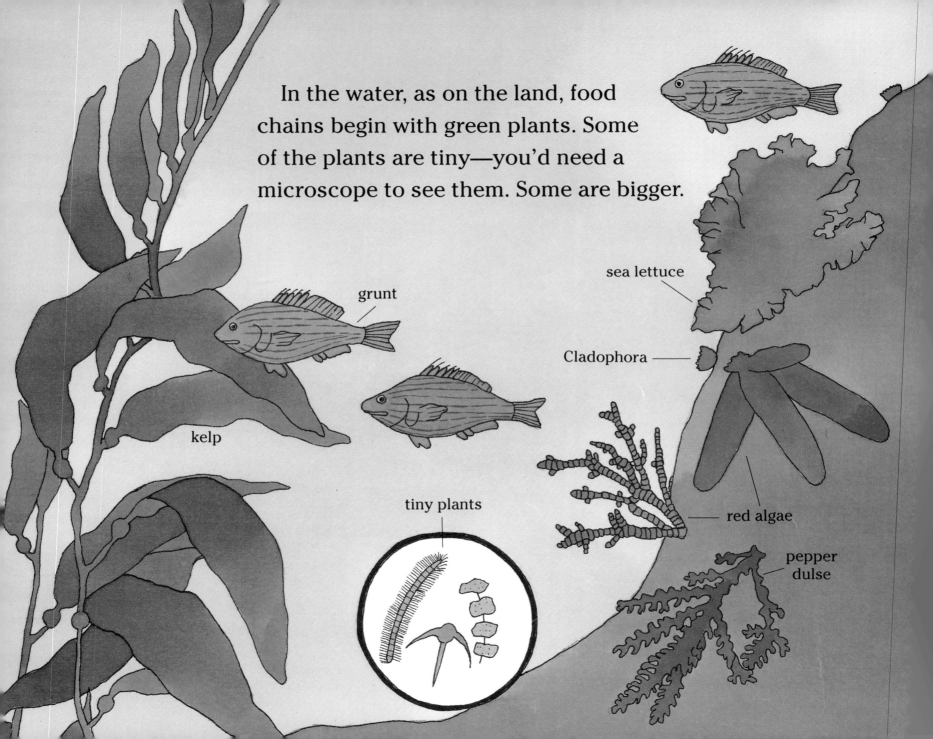

In the water, as on the land, food chains begin with green plants. Some of the plants are tiny—you'd need a microscope to see them. Some are bigger.

grunt

sea lettuce

Cladophora

kelp

tiny plants

red algae

pepper dulse

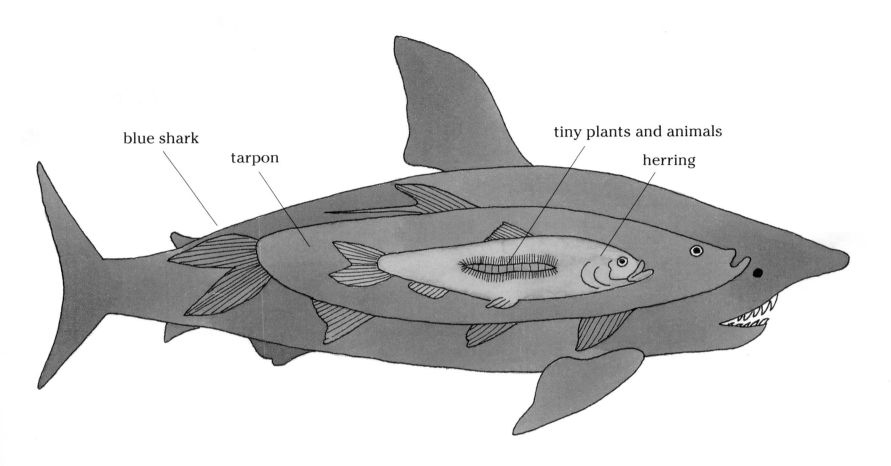

blue shark

tarpon

tiny plants and animals

herring

The green plants are food for many tiny creatures,
which become food for bigger creatures.
Small fish are eaten by bigger fish,
which are eaten by still bigger fish,
which are eaten by even bigger fish.

albacore tuna

The biggest, such as tuna, are at the tops of food chains—unless they are caught by humans. Then one of them may turn up in your tuna-fish sandwich. Both the tuna and you are part of a food chain that began with a tiny green plant.

Food chains are found wherever life is found.

The far south of the world, Antarctica, is icy and bitterly cold for much of the year. But in summer its seas come alive. The water is rich with tiny green plants. They are fed on by tiny animals. And these are fed on by small animals such as krill, which look like shrimp. All these animals and plants are food for bigger animals, such as fish and squid.

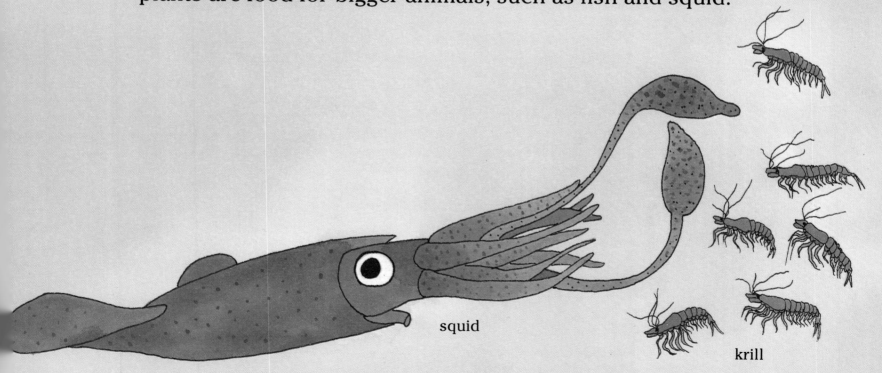

squid

krill

Cape pigeon

petrel

Many other animals come to feast in these waters. There are seals, whales, and dolphins. There are many seabirds, among them penguins.

All the animals are linked to the tiny green plants.

crabeater seal

Adélie penguin

blue whale

killer whale

27

The drawing shows a web of food chains at the far south of the world. The arrows show who eats what. Follow the arrows and find the animals that feed on krill—one of them is the blue whale, the biggest animal on earth. Find the animals that eat animals that eat krill.

Sometimes people talk about catching krill for human food. But what would happen to the food web if fishermen took huge catches of krill each year? To find out, look at the drawing again.

Humans often make changes in food chains and webs. Then they find that one change causes other changes. That was what happened when hunters killed nearly all the Pacific sea otters.

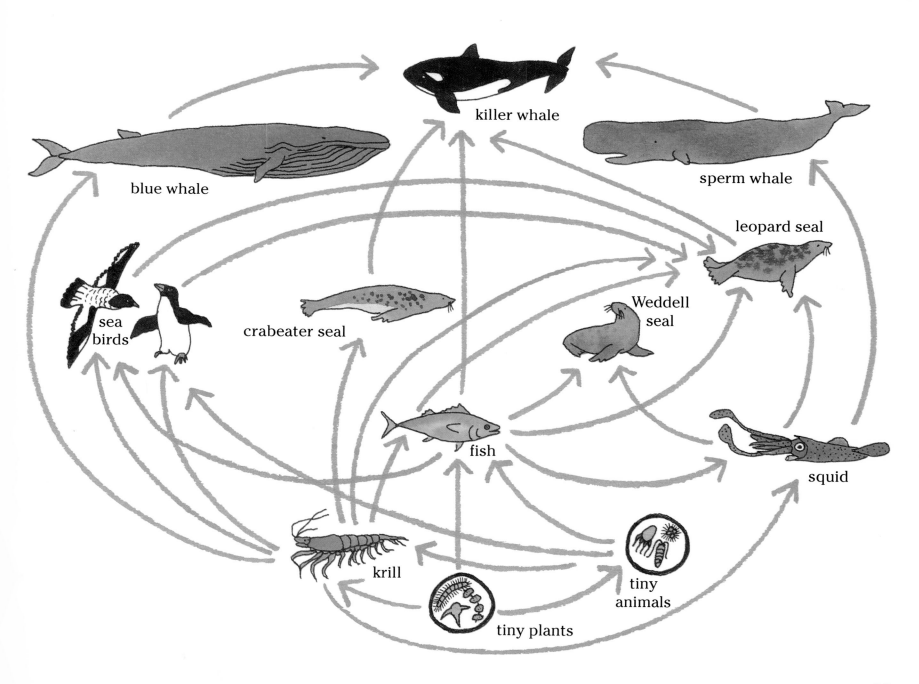

blue whale

killer whale

sperm whale

leopard seal

sea birds

crabeater seal

Weddell seal

fish

squid

krill

tiny animals

tiny plants

29

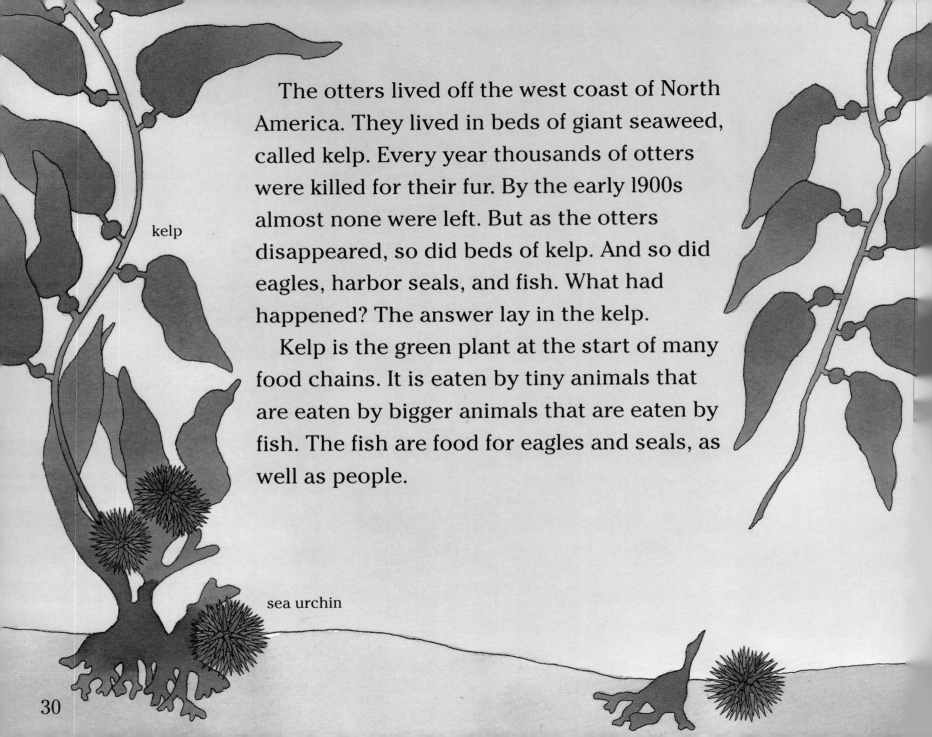

The otters lived off the west coast of North America. They lived in beds of giant seaweed, called kelp. Every year thousands of otters were killed for their fur. By the early 1900s almost none were left. But as the otters disappeared, so did beds of kelp. And so did eagles, harbor seals, and fish. What had happened? The answer lay in the kelp.

Kelp is the green plant at the start of many food chains. It is eaten by tiny animals that are eaten by bigger animals that are eaten by fish. The fish are food for eagles and seals, as well as people.

kelp

sea urchin

30

Kelp is also eaten by spiny animals called sea urchins. In eating, they may cut off stems at the seafloor. The kelp then floats away.

Sea urchins are one of the foods otters like best. But when hunters killed the otters, there was no one to eat the urchins. The urchins destroyed the kelp beds.

Once the hunting stopped, the otters made a comeback. They ate sea urchins, and the kelp began to do well. When the kelp did well, the fish came back—and so did the eagles, seals, and fishermen.

sea otter

seal

31

All over the world, green plants and animals are linked in food chains that branch into food webs. A change in one link is felt up and down that chain. It is felt through the whole web.

And that's one good reason to take care of the earth—to take care of its plants and animals. When we help them, we also help ourselves. We too are part of many food webs.